The Stuff o[illegible]

Matter, Energy, Atoms, and Quantum for the Non-Scientist

by

Irwin Tyler

ISBN: 9798848279191

Ahl Kayn Publications

Spring Valley, New York

HTTP://www.AhlKaynBooksandArtWork.com

TABLE OF CONTENTS

PROLOG 5

INTRODUCTION 6

SECTION 1: THE UNIVERSE IS MADE UP OF "STUFF"..8

Matter is Solid 10

The Universe is Made Up of Hardly Anything 15

Matter has Structure 18

Matter is Almost Empty Space 19

What Holds Matter Together? 21

SUMMARY SECTION 1: THE UNIVERSE IS MADE UP OF "STUFF" 23

SECTION 2: THE UNIVERSE DOESN'T MAKE ANY SENSE 27

Matter is Energy 28

Matter is Waves, Matter is Particles 33

Matter is Made of Fuzzy Things 34

Matter is Fuzzy Space 38

SUMMARY SECTION 2 THE UNIVERSE DOESN'T MAKE ANY SENSE 40

SECTION 3 - THE THEORY OF EVERYTHING................43

THE BIG BANG...44

The General Theory of Relativity 45

What Goes On In Space? 47

What Don't We Know?...49

FROM WATER VAPOR TO WATER ICE 54

Theory and Experiment...58

More Theory...59

More "Particles" and "Forces"? 60

QUARKS...63

BEYOND QUARKS...69

The Standard Model (Grand Unified Theory)...69

String Theory...71

And More 73

UNANSWERED QUESTIONS..74

SUMMARY SECTION 3: THE THEORY OF EVERYTHING ..76

SECTION 4 IS "EVERYTHING" ALL THERE IS? 77

Have we finally figured out how to describe everything that exists? 77

The Standard Model 77

Higgs boson: The 'God Particle' explained..78

The CERN Experiments **80**

And now… **81**

BIBLIOGRAPHY **82**

AUTHOR BIOGRAPHY **86**

A PERSONAL NOTE **87**

INDEX **91**

Your opinion matters:

If it moves you, please leave your thoughts about this book in a Book Review on the website from which you purchased this volume. Both I and other readers would gain much from your sharing.

Thank you.

PROLOG

What is QUANTUM and why does it matter? To state it very simply:

> The discovery of quantum eventually led to the creation of the atomic and hydrogen bombs.
>
> Because of the concept of quantum we have seen the peaceful application of atomic energy in a satellite continuing on its journey through outer space that still sends us its findings although many years have passed.
>
> Computers and cell phones exist because of what we understand about quantum.
>
> Our understanding about how this universe came to be, how it works, and its ultimate fate rests in great measure on our understanding of quantum.

Yet, knowing that quantum here refers to a packet of energy that is fixed in amount/size and can not be divided does not seem like such a remarkable idea. On the contrary, its discovery was a monumental revelation. And that is the reason for this book series.

INTRODUCTION

There is no way anyone can explain quantum theory to the beginner or non-scientist by diving right into the subject. It's just too weird if we look only to our experiences in the world around us to try to validate what scientists understand about quantum theory and quantum physics. Instead, we need to sneak up on it, building our knowledge slowly, recognizing anomalies, trying to explain conundrums and seeming impossibilities, testing and discarding and rebuilding ideas, and validating our theories experimentally, bit by bit until at least some of quantum begins to make sense.

This book seeks to make enough sense to satisfy you without the need for hard physics and mathematics.

You will likely have as many questions at the end of this book series as you have now at its beginning. Well, that's OK. Physicists and mathematicians studying quantum theory in depth have faced the same situation throughout their careers.

Although you will still have questions, the questions at the end will be different from those you had at the beginning because you will have learned a lot. Some will be questions that a future, deeper study of quantum may be able to answer for you. And some may be questions that no one in our lifetime will confidently be able to answer.

Understanding quantum is still a journey in progress.

Irwin Tyler 2022

SECTION 1: The Universe is Made Up of "Stuff"

Our earliest recording of man's inquiry into the nature of "things" comes from Greek philosophical writings of around 1000 BCE. Certainly, there were other societies and earlier societies that also thought about the nature of "things" but it is through the Greek lineage that scientific inquiry and the current understanding of quantum theory springs.

Greek thought was based upon what people could experience, which became the basis for reasoning about the implications of these experiences. They observed permanence and change, stability and movement, unique objects and many objects as a group called by a single name (Fido the dog or all dogs), and they asked: what makes the world what it is? What makes things appear the same yet are different? Things move, but why is the movement of a rock thrown in the air different from the movement of a leaf falling from a tree, or is it? What makes the substance of rock different from the substance of skin? And what is an idea?

In the 5th Century BCE Leucippus of Miletus and his pupil Democritus asked the following question:

What happens if you break a piece of matter in half and in half again. Then do it again and again. Is there a limit to how many breaks you can complete before it can break no further?

They reasoned that at some point there must be an end to the process, where it leaves an unbreakable bit of matter, which Democritus called an atom. Once they reached this conclusion they attempted to explain what these atoms were. They reasoned the following:

1. All matter consists of invisible, undividable particles called atoms.
2. Atoms are indestructible.
3. Atoms are solid but invisible.
4. Atoms are homogenous – an atom is a single substance.
5. There is empty space surrounding all atoms.
6. Atoms differ in size, shape, mass, position, and arrangement.
   - Solids are made of small, pointy atoms.
   - Liquids are made of large, round atoms.
   - Oils are made of very fine, small atoms that can easily slip past each other.

About 100 years later Aristotle, with a different reasoning about the nature of substance, concluded, quite differently, that all substances were made up of varying combinations of air, fire, water and earth. So respected were his opinions and teachings that the ancient world rejected the ideas of Democritus and held to his alternative view of matter.

Because Democritus, in developing his philosophy of reality, proposed the idea that the universe had always been in existence and that there would be no end to it, the Catholic Church debated and rejected atomism and retained as "truth" Aristotle's concept, calling Democritus' theory of the atom an idea that was equivalent to proposing Godlessness. This position held sway until the 1800's, resulting in no significant serious scientific inquiry being pursued throughout this period of 2100 years.

Matter is Solid

In the Age of the Enlightenment, that 100+ year period leading to the 1800's, cultural and intellectual forces in Western Europe began to emphasize reason, analysis, and individualism rather than the authority of church or monarch. This led to a growing number of philosophers and experimenters reviving inquiry into the nature of the universe and its parts.

At the beginning of the 19th Century an English chemist, John Dalton, performed numerous chemical experiments. He was able to demonstrate that matter, as Democritus had proposed, seemed to consist of elementary lumpy particles (atoms). Through his experiments Dalton further deduced and expanded his atomic theory:

- Elements consist of tiny particles called atoms.
- Atoms are solid and indivisible.
- Atoms can be neither created nor destroyed.
- An element is one of a kind (pure) because all atoms of an element are identical.
- All the atoms that make up an element have the same mass (the amount of matter each atom consists of).
- All elements are different from each other because their atoms have different masses.
- A compound is a combination of elements bonded together.
- Chemical reactions involve the rearrangement of combinations of atoms.

Experiments proliferated throughout the 19th Century. English physicist J. J. Thomson made a critical discovery of the electron, a negatively charged particle, in 1897.

It had been observed that certain materials, when rubbed, developed a difference in charge between the two materials. Particles from one of these materials could be "pulled off" and sent through a tube emptied of air. He measured the mass of these particles, which he determined had negative charges, and found that they had a mass very much smaller than the mass of a hydrogen atom.

Thomson theorized that atoms were made up of collections of these negative particles, now known as electrons. But he also observed that atoms had no charge. He then reasoned that atoms had another group of particles with positive charges to balance out the electrons.

Since it was still thought that atoms were solid and indivisible, in 1900 Thomson proposed the Plum Pudding model of the atom. He concluded that atoms are made of positive cores, the pudding, and negatively charged particles, like bits of plum or raisins, embedded within the pudding.

So, by the turn of the 20th century matter was thought to be made up of atoms, solid balls yet formed from two oppositely charged kinds of materials. An atom of each element had a different mass from that of another element, meaning that each element had a different number of these negatively charged particles and positively charged other "stuff". Elements were made up of groups of the "Plum Pudding" atoms stuck together.

Questions still remained. Were the embedded negative particles stationary or moving? If they moved, what paths did they take? Were such paths fixed, random, or changeable?

At the same time, others were performing seemingly unrelated experiments, working with materials and equipment that others were not directly interested in, and making new discoveries. Many resulted in new unanswered questions. Few could realize that some of these other results and ideas would begin to answer their earlier questions.

Matter is Not So Solid

There is no end of amazement at the human brain's ability to propose questions, to question answers, and to devise ways to investigate the world around them.

Henri Becquerel, Pierre and Marie Curie, and Ernest Rutherford each were fascinated by the discovery that certain elements (uranium, thorium, radium, and polonium) spontaneously radiated some kind of energy. It was at first thought to be a form of x-ray. Rutherford's experiments showed that there actually were at least two kinds of radiation. Two that he clearly identified he called alpha rays and beta rays.

Based on earlier experiments with uranium by Becquerel, who used various barriers between uranium and photographic plates to study characteristics of the emitted radiation, Rutherford showed that the alpha rays were less powerful than the beta rays. Somehow, beta rays were able to punch their way through a greater thickness of aluminum than the alpha rays were able to achieve.

Cathode rays had been discovered in 1876. It was known that these rays could be manipulated by electrical devices. Rutherford was able to measure the mass of cathode rays, thus showing that these rays were actually particles. When Rutherford, in his 1897 experiments with beta rays, applied a magnetic field to his experimental apparatus he observed that the beta rays shifted their path. This meant that beta rays were not like x-rays, which would travel a straight path, but were charged particles just like cathode rays. It was soon determined that the beta rays were, in fact, a stream of the recently discovered electron.

Meanwhile, the Curies, experimenting with alpha rays, found that they behaved like a projectile, losing energy as they travelled through thicker and thicker sheets of aluminum. When Rutherford finally used powerful enough magnets in his experiments he was able to observe that alpha rays, like beta rays, could be made to shift their paths. It was known that actual rays could not be affected by magnets. Evidence was now conclusive that alpha rays actually were particles, and much heavier than electrons. This was why he needed more powerful magnets to deflect them. He was now also able to show that these particles had an opposite charge to that of the electron.

Could the Plum Pudding theory of the structure of the atom, a ball of positive particles studded with electrons, explain these new discoveries?

Things were moving fast in the world of experimental physics in the late 19th and early 20th Century years. As we will see, the Plum Pudding theory, as convincing as it was, did not last even a decade.

The Universe is Made Up of Hardly Anything

In the absence of a magnetic field Rutherford bombarded a sheet of gold foil with alpha particles. Most particles passed straight through. Unexpectedly, he observed that an occasional alpha particle changed direction sharply from its original path, sometimes bouncing straight back from the foil. His explanation for this experimental result is as follows:

- Alpha particles are solid.
- Alpha particles can go through gold foil.
- There must be spaces between the atoms of gold foil to allow the alpha particles to pass through.
- Since the atom's "pudding" was reasoned to be positively charged, and the alpha particle was experimentally found to be positively charged, the alpha particles must bounce back because the two positive charges repel each other.
- Since the atom also contains electrons (the raisins or plum bits), if the electron was embedded within the central "pudding" the whole would be electrically neutral. But the experiment shows this not to be the case.
- The only place for the electron to occupy is some small part of the empty space between atoms, and that space must be large enough to allow passage of an alpha particle.

- This means that there is a positive core concentrated into a small region with a large amount of empty space around it, and then some arrangement of electrons further out.

In the 1890s Max Planck had performed a series of heat experiments. As he heated metals they glowed in different colors, and each color radiation had a different amount of energy. In 1900 he showed that atoms absorbed this energy in specific amounts. To illustrate, an atom might absorb 2 units of energy or 3 units of energy but not 2.5 units. He called these units "quanta".

Planck found that the concept of "quanta" could not be explained by any of the laws of physics then known. Despite the experimental demonstrations and mathematical "proofs" that quanta existed, for many years Planck, himself, found the concept

of quanta intellectually illogical, so strong was the pull of classical physics as it had evolved through the 19th Century. That there was empty space in the atom, and that atoms could absorb energy only in specific amounts were just two of the many new revelations about the nature of matter that still had no explanation.

Matter has Structure

The most important outcome of Rutherford's experiments was that they revealed that the atom had a nucleus. By 1909 he had concluded that the bulk of the mass of an atom was in its nucleus. Based on his experiments with alpha particles, Rutherford speculated that the negative electrons orbited a positive center much like the solar system, where the planets orbit the sun.

How much space there might be and the kinds of orbits electrons might travel was still a matter of speculation. He proposed that the orbits of electrons were varied and scattered, creating a kind of cloud of electrons about the nucleus.

By 1920 Rutherford was able to further refine his understanding of the atom's nucleus. He was convinced that there were both positive particles (protons) and a mysterious neutral particle together making up the nucleus.

Since his experiments in shattering the nucleus produced electrons, his thought was that each electron was somehow tightly bound to a proton, creating a neutral particle, a neutron. Thus, his model of the nucleus consisted of a number of protons bound to an equal number of electrons which balanced and cancelled out their charges, plus additional protons that would account for the measured positive charge of the nucleus.

His theory seemed to cover many open questions: he could account for the heavy weight of the nucleus, its positive charge, and his experimental results.

Matter is Almost Empty Space

Earlier, in 1913, Rutherford bombarded nitrogen and several other light elements with alpha particles. He observed that some nuclei could be shattered, with the collision producing fast-moving protons (positively charged particles). But most alpha particles passed through. So, he concluded, the nucleus was mostly empty space.

At this point atoms were known to be neutrally charged, with a proposed nucleus and some kind of arrangement of orbiting electrons and lots of empty space.

But the nucleus needed to be positively charged to balance the negative electrons. The nucleus must, therefore, consist of positively charged particles. These positively charged particles had to be much heavier than the electrons, since they repelled the "large" alpha particles.

The "Plum Pudding" concept of the atom was officially dead. But the picture of the atom still was not complete.

What about all this empty space? Where were the orbiting electrons and why didn't they just fly off into space?

Rutherford's student, Neils Bohr, in 1919 examined earlier experiments and noted logical inconsistencies in proposed theories about the atom. He accepted the generally believed idea that the atom was solar system-like, with electron "planets" orbiting a nucleus "sun", but he uniquely also proposed that the electrons did not merely orbit in a somewhat random "cloud" anywhere outside the nucleus but actually orbited in some fixed arrangement around the nucleus.

In the 1850's scientists, in heating pure elements until they glowed, radiated light (energy) only at specific frequencies/colors. Every element had its specific pattern of frequencies of emitted light. Why this was so could not be explained by the then known principles of physics and theories of the nature of the atom.

Bohr reasoned about these unexplained results and of other unexplainable results in previous experiments with light. He concluded that an orbiting electron absorbed heat energy by shifting its orbit to a higher energy level. When each electron returned to its originally lower energy level it radiated this energy back in the form of light at a specific frequency, representing the difference in the energy contained at each level.

But the supposed randomness of the orbiting electrons did not support all of the experimental results. The specific

frequencies could occur only if there was a fixed arrangement of electron orbits. Physicists now had still another new understanding about the nature of atoms.

What Holds Matter Together?

Experiments of all kinds continued, and in 1930, when Bothe and Becker bombarded beryllium with alpha particles, this produced both the expected charged particles but also an unexpected neutral, uncharged radiation.

Calculations of the energy of all the radiation plus particles released in these collisions showed that the neutral radiation could not be gamma rays, as had been thought at first. Further bombardment experiments by Chadwick in 1932, using different materials, resulted in his determining that this neutral radiation was, in fact, another particle, with a mass close to that of the proton.

By this time it was clear that tiny differences between theoretically calculated energy levels, and particle masses that differed significantly from those measured in carefully constructed experiments, actually were important. These differences hinted that the theories producing the calculations either were incomplete or actually partially or fundamentally wrong. This new understanding would assume greater importance in the future as scientists sought to verify and explain the results of older experiments.

While it was now seen that the atom's nucleus consisted of protons and neutrons packed together, nothing in the experiments changed the understanding of twenty years earlier that the nucleus consisted mostly of empty space.

Nothing as yet could explain why and how the presumed

solar system-like structure of the atom could exist and sustain itself. If negative electrons moved around the positive nucleus, why did electrons not crash into the nucleus? Electrons were known to somehow exist within the nucleus: did these electrons also move? Why do they remain in the nucleus? Why do protons in the nucleus not repel each other instantaneously?

Each new finding seemed to lead to an endless series of new questions and new experiments. And each new finding disclosed

ever tinier "particles" and strange new "things". Three thousand years of serious inquiry still had left the nature of "nature" largely a mystery.

Summary Section 1: The Universe is Made Up of "Stuff"

Our earliest recording of man's inquiry into the nature of "things" comes from Greek philosophical writings of around 1000 BCE. Greek thought was based upon what people could experience.

Well, what happens if you break a piece of matter in half and in half again. Then do it again and again. Is there a limit to how many breaks you can complete before it can break no further? So, Democritus reasoned, all matter consists of invisible, undividable, indestructible particles called atoms. Aristotle thought differently, that all substances were made up of varying combinations of air, fire, water and earth, and his ideas prevailed for 2,100 years.

Early 19th Century experimenters demonstrated that Democritus may have been right. To his conception of matter was added:

- An element is one of a kind (pure) because all atoms of an element are identical.
- All the atoms that make up an element have the same mass (the amount of matter each atom consists of).
- All elements are different from each other because their atoms have different masses.
- A compound is a combination of elements bonded together.
- Chemical reactions involve the rearrangement of combinations of atoms.

Additional experiments by Thomson revealed that atoms consisted of electrons of negative charge and something else of positive charge, with the charges balancing each other. He fit these together into a "plum pudding" model of the atom, with the electrons stuck to the outside of the "positive" pudding.

So, why did some electrons, when hurled at a thin sheet of metal, go right through without seeming to harm the metal? Clearly, there had to be spaces between the atoms making up the metal sheet, and perhaps even within the atom itself.

It turned out that matter was made up more of empty space than of "stuff". The "plum pudding" model of the atom was dead. It was deduced by Neils Bohr that the atom had two parts:

- There is a positive core concentrated into a small region, with a large amount of empty space around it.
- Electrons occupy some part of the space surrounding the core.

This was the "planetary" model of the atom. He concluded that the absorption of energy by electrons orbiting the nucleus did not proceed smoothly but occurred in increasingly larger fixed amounts (packets), step by step. The QUANTUM concept was born.

Experiments in the early 20th Century revealed that the positive core of the atom consisted of at least two particles of similar mass, one positively charged and one without charge. When the nucleus of an atom was smashed electrons emerged. So, the atom seemed to be much more complex than at first thought. This brought out new questions:

- If negative electrons move around the positive nucleus, why do electrons not crash into the nucleus?
- Why do protons in the nucleus not repel each other instantaneously?
- Where are the electrons in the nucleus?
- Are there spaces within the nucleus?

SECTION 2: The Universe Doesn't Make Any Sense

In 1895, before x-rays were discovered, before radioactivity, before the electron was known, science knew only two forces: gravity and electromagnetism; two properties of objects: mass and charge; and an agreed understanding of how these forces acted on objects: Newton's Laws of Motion. Heat and light were well understood. Once the universe was set in motion it would continue forever. It all made sense. This was the "mechanistic" view of the world.

But now new questions were being asked: why is the measured electric charge the intensity that it is? Why do the measured masses of different elements exist in their specific amounts and ratios? Why is the speed of light what it is? What can explain the dark gaps seen in the spectrum of light coming from the sun directed through a prism or thin slit? Why do many metals emit electrons when light shines upon them? Why does light reflect off some surfaces yet become absorbed into others? What really happens when different elements are mixed together?

What, if anything, did these questions have to do with each other? Many questions. Many experiments. Many theories. Many arguments.

Coming into the 20th Century it was clear that different schools of understanding about what "matter" was and how it worked were in collision. Moreover, various experiments in independent areas of research hinted at understandings that seemed to conflict with each other.

Great institutes such as the Carnegie Institution, well-endowed universities such as Princeton, government bureaus such as the U.S. National Bureau of Standards, and corporate research at such giants as AT&T produced a myriad of studies and experiments, and offered conflicting theories that needed better understanding, or might have to be abandoned.

Matter is Energy

1905 was a watershed year, the year that Einstein's theory of "special relativity" showed that matter and energy were equivalent – they could be converted one to the other.

This theory, and his later theory of general relativity, supplanted all earlier assumptions and theories about matter, energy, space, time and gravity.

That matter and energy are equivalent makes no sense in the world we experience every day. It was an idea that never entered the minds of others in his day. Einstein needed to uncover two counter-intuitive principles to explain this equivalency. He discovered the first through several mental experiments whose results could not be explained by classical physics.

(A) The speed of light in a vacuum had been measured with a fair degree of accuracy (186,000 miles per second), and electromagnetic theory describing the behavior of light had already been established. Einstein imagined himself riding on a light wave moving alongside a similar light wave. According to classical theory he and a rider on the other light wave should have moved along together, the motion and speed of himself being equal to the motion and speed of his companion. Since they were each riding a "vehicle" moving at the Speed of Light, their speed relative to each other would be ZERO.

Yet, all then current experiments demonstrated that light always traveled at the same Speed of Light in a vacuum no matter in what direction or speed the measurer was moving. Einstein believed the experimental data. He could only conclude that current reasoning about light was not correct. This led him to imagine another experiment.

(B) A train was coming toward an observer at the "speed of light". At some instant a passenger shone a light beam toward the observer. Classical physics would expect the light beam to approach the observer at a speed that is the sum of its own speed plus the speed of the passenger producing the light beam, here adding up to TWICE the Speed of Light. Yet, experiments showed that light does not measure at a speed that exceeds the Speed of Light.

(C) For an observer moving East very slowly, and an observer moving West very fast, for both of them to measure the speed of the same light beam as exactly the Speed of Light, something must be different concerning the space and time experienced by each observer. The experience of space and time is relative to the situation of the observer, he concluded.

In order for Einstein to make the speed of light constant for all observers under all conditions he had to accept that space and time were not the fixed building blocks understood by classical physics. Why, then, does classical physics seem to explain, with the finest detail, how our world seems to work? Einstein's mathematical equations showed that the kinds of strange "relativistic" results demonstrated in his mental experiments are imperceptible until an object is traveling very close to the Speed of Light.

This first principle contained within Einstein's theory of Special Relativity, concerned with the speed of light, is expressed this way: The speed of light in a vacuum is always the same for all observers under all conditions.

Next, we turn to his second principle, that concerning the laws of physics:

- The Conservation of Energy principle states that in a closed system, where nothing within the system can leak outside and nothing from the outside can enter into it, the total energy contained by two individual about-to-collide objects is the same total after the collision. This is true whether there is a change in position of either object relative to a table below (potential energy) or a change in the speed or direction of motion of either object (kinetic energy). There is also the thermal energy (heat) produced by friction in the collision. The total energy measured within this closed system before and after the collision remains the same.

- The Conservation of Mass principle states that no matter how we rearrange the constituent parts of an object its mass remains the same.

Einstein showed that these classical principles also were relativistic. Imagine this experiment:

One person is in a fixed location. The other, holding an apple in his hand, is accelerating ever closer to the Speed of Light. The traveler finds the apple has a certain weight, and this weight is constant as he moves ever faster. The fixed observer notes that the traveling apple grows heavier as it approaches closer to the Speed of Light, apparently violating the principle of the Conservation of Mass.

Einstein's equations showed that for the moving observer the classical laws of physics hold together. No matter where the traveler was or how fast he was moving, for him classical physics did not change. Einstein proposed that the laws of physics, how things in the world behave, are the same for all observers no matter where they are. But something was different for the other fixed observer. The traveler was in a closed system, and classical physics works there. But for the fixed observer this was an open system in which energy was being added to the traveler and his apple in order to make it accelerate.

Here, too, Einstein's mathematical equations showed that this kind of strange "relativistic" result is imperceptible until an object is traveling very close to the Speed of Light.

What Einstein demonstrated here was that adding energy to a system increases its mass. He also demonstrated the reverse principle, that removing energy from an object reduces its mass. This is expressed in his famous equation $E=MC^2$.

Mass and energy are equivalent to each other and, theoretically, can be converted to one another. We can measure this equivalence through their relationship to the Speed of Light.

It took less than half a century to translate Einstein's revolutionary idea and his mathematics into the reality of controlled nuclear energy and the atomic bomb.

The early years following the 1905 introduction of Special Relativity would seem to have given the notion of matter being solid-like, made up of "things" convertible to energy, a clear lead in scientific thinking. But, other experiments and theories pointed in another direction.

Matter is Waves, Matter is Particles

Now, it had been known for many years that light was a wave. A light beam, "wiggling" in space, could wiggle through two side-by-side slits in a metal sheet.

Experiments with light also showed particle tendencies as well as wave tendencies. Light, having no mass, still could push a vane, a child's toy, and make it rotate.

Here's what we knew: Waves kind-of move around in space, and they have no mass. Particles have mass, they occupy a particular region of space and they follow a specific trajectory as they move. Electrons (particles) rotating around the nucleus can only absorb amounts of energy in specific packets (sizes). But waves are continuous energy "things" that can absorb any amount of additional energy.

Waves and particles are very different yet Einstein showed them to be equivalent and convertible one to the other.

Particles moving at the Speed of Light, according to Special Relativity, are equivalent to pure energy, like light, which is a wave. So, a particle and a wave are, in some sense, the same thing.

It would not be wrong, therefore, to consider particulate matter as waves, and waves as particles. This paradox came to be called the "wave-particle duality".

Wave-particle, of course, makes no sense in the world of everyday experience.

Matter is Made of Fuzzy Things

We have been able to measure the Speed of Light with ever

increasing accuracy. We can compute with extreme accuracy the thrust, trajectory and needed corrections in order to position a spacecraft into orbit around Pluto. So, shouldn't we be able to determine the position and trajectory of an individual electron or proton? Werner Heisenberg, a German physicist, determined that according to quantum principles we can never know both where such a tiny object is and its velocity (its speed and direction of motion). This is called Heisenberg's Uncertainty Principle.

A simplified scientific explanation

Let's remember that Einstein showed that matter and energy are equivalent. His equations made clear that mass moving at the Speed of Light takes the form of pure energy. At rest it is pure matter. Since all mass that we encounter moves at some intermediate speed, matter must somehow be some combination of mass and energy.

So, how can we precisely measure the position, speed and direction of motion of that electron, proton or neutron that we are trying to measure if it isn't the solid particle we usually imagine it to be, if it isn't entirely mass? What part of it would we be measuring? And what would that measurement even mean?

Science thus teaches us that we can't know everything about these fundamental "things" of matter.

A simplified logical explanation

Everything we encounter in life, right down to the atom, seems to us to be solid. Even the makeup of atoms, their electrons, protons and neutrons generally are thought of as particles since they have a rest mass and a size.

Let's now find the position of a moving electron at some instant in time. We need some mechanism that tells us where it is in space. We could shine a light beam on it and by measuring the rebound we can calculate where the electron is. Or we could measure the electromagnetic energy coming from it and calculate where it came from. But there is a problem. No matter how close we get to the surface of the electron particle there still is a distance the light beam or electromagnetic energy must travel to reach our detector. And that takes time.

So, we might be able to know where the electron particle WAS but we can not know where it IS now. Making it even more of a problem, in order to know its speed and direction we would need to take two measurements.

Logic thus teaches us that we can't know everything about a unit, a quantum of matter.

Heisenberg's Uncertainty Principle

Anything that we might hurl at a moving atomic particle in order to make a measurement either is all energy or some combination of mass and energy. Their collision will transfer some amount of energy either to or from the atomic particle. That particle's mass and/or velocity will have changed.

As Heisenberg made clear, at the quantum level the very act of observation changes what is being observed. The best we can do is calculate the probability of the atomic particle being at a particular location plus or minus a little bit, and the atomic particle's velocity, more or less.

We now know that matter is not all mass nor all energy. We also know that matter at the quantum level can behave both like a wave and a solid particle. Finally, we have learned that a particle of matter can never be completely known. Put this all together and we can understand why we must say that matter is made of fuzzy things.

Matter is Fuzzy Space

After Einstein's breakthrough work, a "particle" of matter was understood actually to be its combined mass and energy, with the energy component having no mass of its own.

The energy component of matter was described as including a combination of its momentum (motion), its electric charge, its magnetic field, its potential due to its position in a **force field** such as gravity, its stored energy, as if it were a coiled spring, and maybe other kinds not yet identified.

One school of thought argued that the Uncertainty Principle described a universe which has no definite-ness to it. Everything that can be known about it and its components is merely probability within a degree of certainty. Why? Because mass and energy depend upon how fast the mass is moving, which Einstein showed can be different depending upon the vantage point of the observer.

Neils Bohr and his colleagues, in 1920, argued (the Copenhagen interpretation) that matter and photons (energy) have no actual position and no momentum until the moment they are measured, which, we have now seen, can never be known exactly. They proposed that a particle of matter contains ALL possible natures at the same time, which are determinable only in terms of probabilities. Each time an object is "observed" it forces one of the probabilities (its state) to become exposed. Each observation exposes a different state. This state of existing in all possible states at once is called an object's coherent superposition.

Another proposed view of reality at the atomic level was the "multiple universes interpretation", which says that, according to quantum theory, each possible outcome occurs in its own universe. They exist simultaneously, but in different "places".

Einstein argued that everything is certain even though we can never measure it. One counter to this idea is that, at the quantum level, objects have a wave characteristic, and there is no such thing as the precise physical position of a wave.

Each of these theories and explanations was an attempt to explain how results of experiments on these quantum objects could show erratic, UN-predictable outcomes.

What we were left with in the early 20th Century was objects existing and moving through mostly empty space that, themselves, contained mostly empty space. These were objects whose characteristics could not be known exactly, and even what was believed to be known about these quantum objects would be different to observers measuring them in different experiments and in different places.

In other words, we knew that we had much more to learn about "matter".

Summary Section 2 The Universe Doesn't Make Any Sense

The watershed year of 1905, with the publication of Einstein's Theory of Special Relativity, established a new baseline for the understanding of matter and energy. It was now clear that classical physics in any of its various theories could not explain the behavior of matter on a small scale. "Relativity" provided a better understanding of why experimental results with atomic and subatomic "particles" could be observed as waves as well as particles.

Following Einstein's work, a "particle" of matter was now understood to be its combined mass and energy. The energy component of matter was described as a combination of its momentum (motion), its electric charge, its magnetic field, its potential due to its position in a force field such as gravity, stored energy as if it were a coiled spring, and maybe other kinds of energy as well.

Heisenberg established that for experiments at the scale of atomic dimensions no experiment could ever be complete. He explained that either the position or velocity (speed and direction of motion) of these electrons, protons, and neutrons could be determined, but not both at the same time.

So, in the early years of the 20th Century scientists believed that objects containing mostly empty space existed and moved through mostly empty space. This could not yet be explained. Moreover, despite new information from experiments and ground-breaking new theories, many previously unanswered questions remained unanswered, such as:

- If negative electrons move around the positive nucleus, why do electrons not crash into the nucleus?
- Why do protons in the nucleus not repel each other instantaneously?
- Where are the electrons in the nucleus?
- Are there spaces within the nucleus?

New questions also began to arise:

- What powers the stars?
- Where did all this matter come from?
- If matter decays, why is there still so much of it?
- If the universe started with mixtures of positive and negative charges, why haven't they attracted each other and completely annihilated themselves?
- Will the universe last forever?

SECTION 3 - The Theory of Everything

While theoretical scientists and experimenters mainly pursued knowledge about ever smaller "things", some began to look at the other extreme. Here, the major force affecting massive objects was gravity. This force seemed peculiar in that it existed everywhere yet seemed to be irrelevant to the extremely small "things".

The electromagnetic theory was proposed by Maxwell in 1873, explaining the relationship between electricity and magnetism, in effect uniting the individual forces of electric charge and magnetism into a single fundamental force that under the proper circumstances splits into the two component forces. This eventually led theoreticians to talk about the possibility that a single theory could be developed that would not only explain all kinds of matter and all kinds of energy and waves and particles, all at both the largest scales and the smallest, but would show that all these forces were sort-of subdivisions of one single grand force.

This "theory of everything" has proven to be elusive. The two major accepted ideas, Einstein's Theory of General Relativity and the Quantum Field Theory, seemed to work only at opposite extremes of the size and mass scale. Neither could explain the opposite extreme.

Within their specific domains each theory showed itself to be highly accurate and predictive. But the universe actually seems to be a unified whole. So, how could both theories be correct? And how does gravity, the outlier, fit in?

The Big Bang

Where did all the known forces we experience come from?

Some said everything that is, "things", energy and forces, always was and always will be. We observe in space that planets and galaxies all move about. But this theory says that on the whole everything is balanced out and always was balanced out.

Others said that everything somehow came into being. If so, did everything come into being just the way we find it now? If not, what gave us these "things", energy and forces?

Creating theories, in itself, is not very helpful to us. We need something that reveals which is correct, if in fact one of these really is correct.

How can we find out? A lot of this pursuit for answers came out of another of Einstein's revolutionary ideas, the General Theory of Relativity.

The General Theory of Relativity

Einstein imagined a problem: If a man jumps off a building he will feel nothing, no force acting on his body. Therefore, he should fall at a constant speed. But we know that he actually will accelerate. Something we call gravity is acting like a force on him, making his fall speed up. How does this work?

Einstein reasoned that since the man moves through space, and his movement happens over time, something must be happening to space and time even though it doesn't act as a force he can feel or measure. He imagined that space and time became warped by objects, that is, "things" having mass, but that the mass can not feel any force when it moves through space. Since both space and time are affected he called that space-time.

His Special Theory of Relativity showed that because of the

mass of the earth, time at the top of the building would move slower than time at the bottom of the building. We always think of the two times as the same, but since one time really is longer than the other there must be a curving of the time in order for the two times to fit together.

How is that again? Take a long 2x4 piece of wood and start to bend it (assuming you are strong enough). One side curves in, trying to shorten, and the other side tries to lengthen. That's because the wood is trying to maintain the same length on both sides. Eventually it will break.

Because of nearby mass, space-time tries to curve and shorten in one location and lengthen in another. It doesn't break like the wood, it just warps.

So, what is gravity? It's a warping of space-time by a mass. Let us say that two massive objects, for instance planets, approach each other, each warping space. It looks sort of like when a marble rests on a rubber mat it sinks in and distorts the surface. In our example each mass distorts space-time, with the two distortions touching each other lightly from a great distance. Each mass slides into the combined distortion toward the other mass. As the masses get nearer each other the combined distortions get bigger and the masses slide faster toward each other. To viewers looking at this from space it appears that the two masses are being drawn toward each other by an invisible force that gets stronger as the masses approach each other, and they give this invisible "force" the name: gravity. But, of course, there's no force, just the warping of space-time by two masses.

What Goes On In Space?

Just as the whistle from a moving train changes pitch as it approaches and then leaves an observer, light does a similar thing. Light from a star moving away from Earth will be a different color from light that is coming from a star moving toward Earth. This is called the redshift.

Astronomers in the 1920s analyzed redshifts from many stars and galaxies and concluded that, in general, distant galaxies were drifting apart from each other. The universe was observed to be expanding. Well, if the universe is expanding there must be a point from which the expansion began. If we "wind" the universe backwards we squeeze all of this "stuff", mass and energy, into an ever smaller space. What happens as masses get closer to each other? They increase the warping of space and squeeze ever closer, faster. The more squeezed all this mass becomes the hotter it all gets. As mass speeds up and approaches the speed of light, more and more mass converts to energy, as Einstein's Theory of Special Relativity showed.

Mass squeezes tighter, energy increases, and space becomes smaller. At some point, according to this analysis, all mass is energy of some kind, and space shrinks down to ZERO, what is called a singularity. How does the universe and all of its energy shrink down to nothing? We still don't know.

But thc Big Bang Theory states that there was a singularity, and that from this singularity came an explosion of pure energy, creating an infinitesimal space-time. As space-time expanded , the energy within it began to cool, and a variety of different energies condensed out. When some of these energies cooled enough they began to convert into matter/mass, first as subatomic particles, then as atoms and molecules. These massive clouds of matter eventually coalesced through the mechanism we conveniently call gravity, to form stars, galaxies, planets and, eventually, us. Using the best understanding of the laws of physics that we have, the Big Bang started 13.5 billion years ago.

What Don't We Know?

We have deduced that early in the expansion and cooling from the singularity a form of energy we call electromagnetic energy condensed out of some other, more intense energy. With further cooling of space the electromagnetic energy split out into electric charge and magnetism. But there are other energies and masses, with many of them discovered toward the end of the 20th Century. What did these condense out of? What were these more intense energies?

Quantum mechanics (the science of the very small) still can not explain the earliest moments of the Big Bang. It has no mechanism for describing the coalescing of atomic matter/energy into significant mass and the eventual evolving of stars and galaxies

.Gravity (describing the very large) can not explain any of the early expansion of the universe, before significant amounts of matter had formed.

If the singularity and rapid expansion of space is understood correctly, there must have been a very early time when there was

quantum-gravity, an energy that had the nature of acting both on large masses and tiny masses and energy. Earlier, there was a time when there was a unified force, quantum-gravity-electromagnetism, and there was something even earlier that we have not yet identified and can not describe. Several new theories, not yet mature enough to answer these questions, are being explored. These include superstring and supermembrane theory, and supersymmetric matrix theory.

Interestingly, one of the ideas predicted by General Relativity, the "wormhole", a form of warped space-time that was further developed in 1935 as the Einstein-Rosen Bridge, and popularized in the Star Trek series, is being investigated in some of these newer theories. Wormholes are described as tunnels through space-time linking distant parts of space. How to navigate a wormhole and what might be at the other end of it is not yet well thought out.

What else did General Relativity predict? Black holes, for one thing. As masses "fall" into each other and become one they do two things. First, they eventually squeeze into an ever smaller space which contains more and more mass. Then, this increasing mass can reach a stage that warps space-time so severely that any matter or energy, including light, that reaches the surface of the squeezed-in mass and is not energetic enough to continue on and leave the presence of the mass becomes trapped. Since light enters but does not leave, it appears black. That's why the mass is called a black hole. The surface position where some mass or energy can enter but not leave is called the event horizon.

There's even more. Suppose an astronomer looks out at a distant massive star. This star is directly in line with another star much further away, shielding it from view. Or does it? The massive star, according to Einstein's theory, causes a severe warping of space all around it. As light from the distant star nears the massive star it travels within the distortion, with some of the light taking a curved path around the massive planet and then continuing on its way to Earth. So, the hidden star can be seen. Not only that, the light, by traveling a curved path, behaves much like light curving its way through a magnifying class. The effect is the same: the distant star not only is no longer hidden, its image is magnified. This is called "gravitational lensing". That the path of light could be bent was proved by observations of the 1919 solar eclipse. However, it was not until 1979 that gravitational lensing was first observed, with the actual measurements confirming Einstein's calculations.

General Relativity predicted that massively colliding objects would severely disturb space-time, causing it to ripple out, something like the ripples that form in the water when a stone is thrown into a lake. There are strong experimental hints of these gravity wave ripples but there has as yet been no direct detection of gravity waves.

Based on General Relativity, theorists, including Einstein, attempted to determine the fate of the universe. Yes, the universe is seen to be expanding but will the expansion eventually stop, then turn back on itself and race back toward the singularity and create a new Big Bang, again and again? Or, will the expansion continue indefinitely, meaning there was one and will be only one Big Bang.

In fact, recent observations indicate that the expansion of the universe is speeding up. Is there more mass and energy "out there" that warps space-time and produces the accelerating expansion we see? If it's there, why can't we see or detect it? Scientists can calculate the amount and location of what this "stuff" should be in order to produce the effects we see and measure. On the other hand, some scientists believe this "dark matter" and "dark energy" are different from the matter and energy we ordinarily encounter. We just don't know at this time.

As the universe continues to expand, energies and energy sources out at its fringe move further and further away from us. It may well be that there is something "out there" so far away that its light and other energies, traveling at the speed of light, or slower, has not yet had enough time to reach us.

We don't yet know what we don't know, and there's much that we can not yet see or measure – and maybe we never will.

From Water Vapor to Water to Ice

Quantum Theory, or Quantum Mechanics, deals with the extremely small and extremely light, these being, early in the 20th Century, molecules, atoms, and sub-atomic particles such as electrons, protons, and light. Also fairly well understood were three ordinary states of matter: gas, liquid, and solid. Clouds of water vapor, liquid water, and ice are the easily recognized states of water, for example.

It was known that many substances could be found naturally or created experimentally in two or all three of these forms, and they could be induced to change from one form to another. Yet, in each form, using water as the example, each state consisted of the same molecules, each molecule consisting of two atoms of hydrogen and one atom of oxygen linked together unchanged. How could this be explained? Einstein's theories did not seem capable of even describing these states of matter, let alone explaining how these could be.

Matter, it was understood at the time, was made up of things, atoms and molecules, in constant motion, with lots of space between them. At the level of atoms and molecules this motion was determined to be random, with the spaces between them constantly changing. It was determined that the average amount of energy of the atoms or molecules in a mass of substance made a difference. Also making a difference was the average distance between the atoms and molecules, as well as how the atoms and molecules were arranged with respect to each other. The combination of these conditions seemed to govern in which of these three states a substance would present itself.

We are all familiar with these states:

Solids

Molecules of solids are tightly packed, vibrating about a fixed position.

Solids have a <u>definite</u> shape and a <u>definite</u> volume.

Liquids

Liquids have a <u>definite</u> volume but an <u>indefinite</u> shape, which varies according to the container confining them.

Gases (Vapors)

Molecules of gases are very far apart and move freely.

Gases have an indefinite shape and an indefinite volume.

Here is how the different states of matter can be described as relating to one another:

Adding energy to a substance increases the motion of its particles – this motion is called heat. More heat makes the molecules vibrate more. As they move about with less and less limit, at some point the substance changes from solid to liquid. Further heat eventually causes it to change to vapor. When that heat is given up vibrations are lessened and the distance between molecules is reduced, eventually causing the vapor to condense to water. Further loss of heat causes the water to freeze to a solid.

It also is possible for a state to change directly from solid to gas, or from gas to solid. Dry ice (CO2) does this naturally. A significant change in pressure can force other substances as well to bypass the liquid state.

A fourth state of matter has been found to exist, and that requires significantly more energy in order to occur:

Plasmas

Suppose more energy is provided to a gas. At some point electrons moving around the atomic nucleus break free, creating a gas-like substance made up of reasonably freely moving, disassociated positively and negatively charged particles. This is a state of matter called plasma. Lightening can produce channels of plasma. Plasma also can be found in interstellar gases, as well as being the matter inside stars.

Extreme pressures can force atoms to squeeze closer together, creating a unique state of solid. A characteristic of such a solid might be a harder substance than normal, or a substance with an atypical temperature at which it will melt.

What we are experiencing here are the most basic and obvious characteristics that 19th and very early 20th Century physics and chemistry had been able to measure, describe

and theorize about. As more and more experiments were conceived and refined, as measurements became more precise, and as ever more sophisticated forms of mathematics were applied to try to explain typical as well as unexpected results of experiments, attention turned to the most difficult problems: explaining what the essence of the atom was and how the atom worked.

Theory and Experiment

When theoreticians and experimenters began to look inside atomic objects they found that even with the most sensitive measurements results and theories did not always match closely enough. In some cases theory proposed a force, an energy, or a "particle" that was not yet observed to exist.

In 1928, Paul Dirac produced equations whose solutions predicted a strange, unthinkable result: a positively charged electron. By 1932, from Carl Anderson's high energy experiments using cosmic rays, this anti-electron (positron) was found, proving Dirac's equations.

It was then logically proposed that for each variety of matter there should exist a corresponding "opposite", an antimatter. By 1955 scientists had found many such particles of antimatter, including the antiproton and the antineutron. They found that matter and antimatter annihilate each other when they meet. If both matter and antimatter were produced in the early universe, which current theory proposes, why does the universe consist mostly of regular matter and not an equal amount of matter and antimatter? More to the point, how can the universe exist at all? Why wasn't all matter and antimatter annihilated? Physicists can not yet explain this.

More Theory

Dirac addressed the wave nature of atomic particles, particularly in relation to the electron. He proposed that the electron wave had four components, one of which was spin, much like the spinning of the earth. His equations made quantum mechanics and special relativity consistent with each other by describing the behavior of a relativistically moving

electron, that is, an electron moving at speeds close to the speed of light. His equations predicted the existence of the soon-to-be-discovered positron. Other states of the electron were predicted and were taken seriously now even before their discovery.

Moreover, theories and proposed solutions by noted scientists Schrödinger and Pauli that seemed to be incomplete or which failed certain tests were soon shown to be accurate when they were considered as subsets of Dirac's more comprehensive solutions.

More "Particles" and "Forces"?

So, atomic particles were waves, there were wave/particles that were mirror images of each other, and these wave/particles had at least four component characteristics. With all this new knowledge, nothing yet explained how the protons of the atomic nucleus held together. Since there were new "particles" to be discovered, maybe there also were new "forces" waiting to be discovered. Somewhere in all of this there might be an answer to the question.

When the proton and neutron nucleus was bombarded with atomic particles sometimes electrons came out, and sometimes light energy came out. Something must be holding all of these in place before bombardment. This something, presumed to be a force, also must be stronger than the electromagnetic force that should have been pulling the negatively charged outer electrons into the positively charged nucleus. And since this force had not yet been detected it must be acting only across very small distances.

In 1932 Werner Heisenberg proposed that photons of light, shooting rapidly back and forth between the neutrons, electrons, and protons was how electromagnetic forces acted between the particles. Japanese physicist Hideki Yukawa, in 1935, proposed that the assumed force holding the nucleus together acted in the same way. Based on the amount of energy he calculated was needed to hold nucleons (particles in the nucleus) together, he determined that this force had to be carried in a massive particle. It was so hard to detect because it existed at this extreme strength for a very short time and then disappeared, but constantly appearing and disappearing.

It was not until 1947 that this unusual particle, the pion, was detected. The force it carried was called the "strong force".

In relatively rapid succession, thanks to the development of powerful particle accelerators, the results of atomic and subatomic particles smashing together at very high energies/speeds revealed many more new particles.

Think about this for a moment: why should an atom of radium or uranium, sitting quietly, suddenly split of its own accord? If the pion holds the nucleus together, and photons transmit only the electromagnetic force, what causes the nucleus to split (atomic fission)? Among the new particles discovered, it turns out, are those that control how atomic radiation occurs (the W+, W-, and Z particles). These particles carry forms of the "weak force". The "weak force" exhibits its field strength over distances thousands of times shorter than the "strong" force.

QUARKS

Let's go back for a moment to the 5th Century BCE, when Miletus and his pupil Democritus asked the following question:

What happens if you break a piece of matter in half and in half again. Then do it again and again. Is there a limit to how many breaks you can complete before it can break no further?

They could go no further back than to the concept of the atom. The atom eventually was seen to be made up of yet smaller parts: electron, proton and neutron. Suddenly, in the mid-20th Century, a whole gallery of new particles and energies began to appear. If we continue the reasoning process of Miletus and his pupil Democritus then physicists had to ask: what are these particles like if we cut/break them further? This was especially important because these particles were carrying strange forces that then current theory had not predicted.

In 1960, Murray Gell-Mann and Yuval Ne'man each devised a method for classifying all the particles then known. This provided physicists a new way of thinking about and relating these particles to each other. In 1964 it led Murray Gell-Mann to propose that there was another level of "particle" down from the known particles. He called them "quarks" (from a phrase in the James Joyce book, Finnegans Wake).

In Gell-Mann's thinking, quarks came in three separate "flavors", each also having a mirror image anti-quark. He named the three flavors/characteristics, fancifully: "up", "down", and "strange". Within 20 years the developing theory predicted three more quarks: "charm," "bottom" (or "beauty"), and "top" (or "truth"), each with a corresponding anti-quark (a top anti-quark is not the "anti" bottom quark). So, now theory predicted that the most elementary particles known were actually made up of combinations of 6 quarks and 6 anti-quarks.

By combining the quarks in various ways the theory of the quark could now explain the existence of several particles, including the protons and neutrons in the nucleus of the atom.

But something has to hold the quarks together. And so, the "gluon" was proposed to do this job.

This new understanding, as yet unproved at the time, revealed that quarks do not exist by themselves. They must combine, and they can combine only in pairs (mesons) and triplets (baryons). These are members of a family of particles called "hadrons".

- Protons and neutrons each are "baryons", made up of 3 quarks. These particles are subject to the "strong" force.
    - The proton, for instance, is made up of two "up" quarks and one "down" quark.
    - The neutron is made up of two "down" quarks and one "up" quark.
- Mesons are made up of one quark and one anti-quark bound together by the "strong" force.
    - The pion is now understood to be a meson.

Another class of fundamental building blocks of matter was predicted which was not based on quarks. The theory showed that the electron and the neutrino (an uncharged particle) are "leptons", which are subject to the "weak" force.

The Standard Model is the most current theory describing the fundamental particles, forces, and mass entities. It has extended enormously the number of fundamental building blocks of matter we can expect to identify.

By the beginning of the 21st Century most of these particles, anti-particles and forces had been experimentally detected. These are all considered quantum entities. That is, each contains or is of the nature of a single packet of "flavors" which can not be altered or subdivided. These include:

- Six "flavors" of quarks: up, down, bottom, top, strange, and charm;

- Six types of leptons: electron, electron neutrino, muon, muon neutrino, tau, tau neutrino;

- Twelve guage bosons (force carriers), the photon of electromagnetism, the three W and Z bosons of the weak force, and the eight gluons of the strong force;

- The Higgs boson (helps to explain what mass is).

The Standard Model predicts these and many other particles. Not all have been detected in natural substances or even in experiments:

The "graviton", a carrier of gravity "force", has been proposed as a way to explain the nature of a "black hole".

The Theory of Supersymmetry requires that Fermions (all six quarks and all six leptons viewed as a particle family) have anti-partners.

Various combinations of quarks and anti-quarks have been predicted, including some containing combinations of more than 3 quarks and/or anti-quarks.

Quantum experiments sometimes reveal very strange things. Quantum entanglement, for instance, is measurable but unexplainable. In this real experiment a photon can be made to split into two separate photons and sent in two different directions, reaching hundreds of miles apart. If photon "A" is given spin "up" we will measure photon "B" as having spin "down". If photon "A" is now given spin "down" we will measure photon "B" as having spin "up". The two photons, though separate, are "entangled".

Our current theories can not tell us everything about quarks. Only experiment can give us a clue. The problem here is that we are dealing with ever smaller particles: the atom is about one million times smaller than a human hair, and the proton is 100,000 times smaller than the atom. We are close to the limit with current technology as to the smallest object we can measure (a Planck length, about 1.616 x 10-35 meter (6.362 x 10-34 inch), the size of quantum foam (or space-time foam) – whatever that is). That limits what we can learn from quantum experiments.

So, we are left with possibly incomplete thoughts about what the quark is and how it behaves. One such bizarre thought is that the quark is a point-like particle that occupies no space. Well, imagine that if you can!

BEYOND QUARKS

If you thought we were finished, I'm sorry but the train has not reached its destination yet. It is not at all unexpected that some scientists have wondered if quarks, leptons, and bosons are made of something even smaller. Others believe that we have reached the limit and have determined all the elementary particles that can exist.

The Theory of Everything (ToE) becomes the last step, describing ALL particles and energies. It describes what they are in every detail, and how they interact, both at the macro level (the everyday level we live in) and the micro level (where the fundamental particles and energies "do their thing").

The Standard Model (Grand Unified Theory)

The Standard Model, the most developed ToE, encompasses everything we have discussed about the identified and theorized fundamental particles. If we accept the completeness of the Standard Model then all particles have been identified and it remains only to complete the experiments that will confirm their existence. Some of these particles, not yet confirmed in experiments, may have existed at the Big Bang or very shortly after, at energy levels far beyond what our experiments can create today. Most of these particles would have disappeared as the nascent universe cooled. Of those that remain or that we have theorized about, Quantum Mechanics describes accurately these individual fundamental particles. Moreover, Quantum Field Theory very accurately describes their interactions. So, what's left?

For one thing, the Standard Model, and Quantum Mechanics, work at the relativistic levels (as described by Einstein's works). At the levels of large mass, slow movements, and gravity these theories break down.

For another, at the moment of the Big Bang various of their equations predict infinities, about which we can not make much sense. What is infinite mass? What is infinitely small space? What does time mean at infinity?

Gravity, itself, presents a problem. The electromagnetic, weak, and strong forces all act in a given space-time. That is, at a specific time and in a specific location, and in relation to whatever else may be near that location. All characteristics of these three forces can be measured. In contrast, as Einstein proposed in his General Theory of Relativity, gravitation is not an interaction that takes place **IN** a given place and ***at*** a given time, since gravitational forces are a manifestation of the curvature ***of*** space-time itself.

If everything at a fundamental level is quantized, as all major theories propose, then what does a quantized space-time mean? There is no reasonable current theory that provides an answer.

But if gravity is fundamental but not quantized then quantumization is not necessarily a complete blueprint describing the composition of the universe and its history. If true, this can have a profound impact on current theories of "everything".

On the other hand, one answer to this question is to reject that gravity is fundamental but that it results from certain interactions. This is not yet a well-formed theory but, if true, it has profound implications to current theories of "everything".

String Theory

String Theory, followed by modifications, such as Superstring/M (M-Theory), postulate that everything in the universe consists of vibrating strings. It is impossible at this point to even discuss what these strings are. At any rate, the theory states that the way these strings vibrate determines what we observe and experience today as the fundamental particles and energies.

According to some scientists, at the Big Bang a single fundamental force turned into four forces at the beginning of the expansion of the universe from its creation. The fundamental nature of everything was vibrating strings. As the universe continued to expand and cool these strings changed their vibrational patterns in many ways, thereafter appearing to us in ways we recognize as mass, force, and electric charge.

A theory born mainly in the 1970's, what gave String Theory some traction was that it had an explanation, although at a simplistic level, of the predicted graviton as well as being able to address all other fundamental particles.

String Theory is not yet as advanced as ToE, although it seems to be able to overcome the difficulties ToE has encountered. It does this in an unexpected way.

We live in, that is we are able to take notice of, four dimensions: length, breadth, height, and time. String Theory proposes that we actually live in many more dimensions. Some say ten dimensions and some say eleven, but we can detect none beyond the four.

In addition, the proposed fundamental strings can take many shapes and may be closed loops or open strings. Following this theory to its limits, there can be multiple universes and an infinite number of particles. Our own universe, according to one extension of String Theory, is that the laws of physics that govern our universe will be different from the multitude of other sets of laws that govern these other universes.

Could a worm hole connect different universes? Can we ever detect these universes? Do they exist beyond the edge of our expanding universe? What happens if these universes with different laws of physics touch?

And More

There are other ToE candidates, all much less developed and not at all accepted by the majority of scientists. Nonetheless, they each seem to answer some of the mysteries not yet completely solved by the major ToE candidates:

- Constructor Theory is an increasingly popular new theory with a unique approach that has recently been proposed by David Deutsch, a quantum physicist at the University of Oxford, in 2012. This theory focuses not on the objects that make up our universe but on the laws that govern their behavior.

- Other theories which are supported by some noted scientists focus on portions of the Standard Model that are incomplete, or that are not addressed at this time. Without getting into what these are, let me name just a few:

- Theory of causal fermion systems
- Causal sets
- Garrett Lisi's E8 proposal
- Christoph Schiller's Strand Model

Unanswered questions

There are many "WHAT" questions that these theories have not yet satisfactorily answered:

- What happened before the Big Bang?
- What is the original Big Bang energy?
- What happened to all the positrons and anti-protons and anti-neutrons in the universe?
- What is dark energy and dark matter?
- If the Big Bang created the universe, what is beyond the edge of the universe?
- If we can roll back time, in our equations, to the moment of the Big Bang, what was there before that time?

- If particles and anti-particles annihilate each other, how do mesons continue to exist?
- What is the string of "String Theory"?

There are many "WHY" questions that remain unanswered by any of these theories:

- Why is the strength of the electric charge of the electron exactly what it is and no other?
- Why is the neutron in the nucleus necessary?
- Why did the early universe suddenly increase its rate of expansion?
- Why are there exactly six quarks and six leptons?
- Why is empty space not empty?
- Why is the speed of light constant?
- Since several ToE equations predict that there should be multitudes of single-pole magnetic particles (only NORTH or only SOUTH – magnetic monopoles, if you wish) floating through the universe, why have none ever been detected?

Summary Section 3: The Theory of Everything

This "theory of everything" has proven to be elusive. The two major accepted ideas, Einstein's Theory of General Relativity and the Quantum Field Theory, seemed to work only at opposite extremes of the size and mass scale. Neither could explain the opposite extreme.

Electrons, protons, neutrons, particles and waves, relativity, mass, gravity, the Big Bang, black holes, quantum, the expanding universe, matter and anti-matter, quarks and anti-quarks, string theory, wormholes, time. As the nature of each building block was proposed and experiments were developed, refined, and completed scientists continued to find that they understood less about how the universe and its parts work than they had believed they were just about ready to explain.

So, alternative theories of "everything" were proposed. There remain open questions that the Standard Model can not reliably answer, and there are experiments yet undone or which, when completed, do not reliably confirm the theory in every detail.

SECTION 4 Is "everything" all there is?

Have we finally figured out how to describe everything that exists?

The Standard Model

The Standard Model is a set of equations and understandings that takes into account everything we have observed and measured about our universe, from the makeup and interactions among the "things" of which the universe is made, to their interactions at the enormous levels of the universe itself.

The Standard Model explains the basic laws that govern the universe and its fundamental forces. There are four fundamental forces: strong, and weak nuclear forces, electromagnetism, and gravity. The strong force holds nuclei together, although some nuclei decay radioactively because they are unstable, causing them to slowly release energy through the emission of particles. This process occurs due to weak forces.

But, in the 1960s, there remained unanswered questions, such as, how do the protons in the atom's nucleus not repel each other, thereby making the existence of the atom impossible? The Standard Model of particle physics revealed, theoretically, a solution.

Higgs boson: The 'God Particle' explained

The Higgs boson is the proposed fundamental force-carrying particle of the Higgs field, first proposed in the mid-nineteen-sixties by Peter Higgs. It is responsible for granting other particles their mass.

Apparently, this expression of mass is strong enough to overcome the nature of the proton's electrical charge that otherwise causes them to repel each other.

Compared to the electron's mass, the boson's mass is theorized to be enormous. It has no charge and zero spin. The Higgs Boson is the only elementary particle with no spin.

A boson is a "force carrier" particle that comes into play when particles interact with each other, with a boson exchanged during this interaction.

At the quantum level, particles are waves and waves are particles. If this boson exists as the theory proposes, how do we find it and measure it?

How do we detect this particle in a wave field, this wave that has no definitive position in space? But it must be found (proven to exist), for without the Higgs boson and the Higgs field in general no particles would have mass. That means no **stars**, no planets, and no us.

In 1964, researchers began to use quantum field theory to study the weak nuclear force — which determines the atomic decay of elements as protons transform to neutrons . What were needed in the equations were not one but two bosons, the W and Z boson force carriers.

Higgs and his associate proposed that when the universe was born it was filled with the Higgs field in a symmetrical, but unstable state. The field quickly found a stable configuration, in the process giving rise to the Brout-Englert-Higgs mechanism which grants mass to the W and Z bosons as well as to many other fundamental particles.

Without the Higgs field and the Brout-Englert-Higgs mechanism, all fundamental particles would race around the universe at the speed of light. Particles that interact with the Higgs field have mass. The stronger that interaction the more strongly they are perceived to have greater masses. But the photon, light, does not interact with the Higgs field, thus it has no mass.

Well, there was now a well-established theory. Needed next was to discover evidence of the Higgs field by detecting its force-carrying particle. It turns out that it would require 50 years, the largest experiment, and most sophisticated machine in human history.

Scientists needed to determine if it precisely matches the predictions of the Standard Model of particle physics. If not, it likely would lead to new scientific discoveries and a different Theory of Everything.

The CERN Experiments

In 1983, experiments at the European Organization for Nuclear Research (CERN) found the first solid, although tentative, evidence for the existence of the W boson. But, jubilation came too soon.

Experiments at the Collider Detector at Fermilab (CDF) had caused trillions of protons and antiprotons to collide with each other at high energies, in order to create W bosons, at a rate of one per 10 million collisions. In 2011, data was reviewed and recalculated from an old experiment. It revealed the most accurate measurement of the mass of the "W boson" particle, showing it weighs more than the Standard Model suggests it should.

The new measurement suggested that the Standard Model was incomplete.

And now…

Later experiments and in-depth research brought more accuracy to experimental measurements. It now shows the Higgs boson particle continuing to behave in line with the theory. Still more studies are needed to verify with even higher precision that the discovered particle has **all** of the properties predicted, or whether, as described by some theories, more Higgs bosons exist, or even a new particle.

Truly, the real joy of science is that there still are many questions to be answered.

BIBLIOGRAPHY

http://www.nobeliefs.com/atom.htm

http://the-history-of-the-atom.wikispaces.com/Democritus

Philosophy and Philosophers: An Introduction to Western Philosophy

By John Shand, Published 2014 by Routledge, London and New York

http://www.chemteam.info/Radioactivity/Disc-of-Alpha&Beta.html

http://www.chemteam.info/Radioactivity/Disc-Alpha&Beta-Particles.html

https://en.wikipedia.org/wiki/Atomic_orbital

http://hyperphysics.phy-astr.gsu.edu/hbase/Particles/neutrondis.html

The Teaching Company: Science in the Twentieth Century

Lecturer: Steven L. Goldman

https://en.wikipedia.org/wiki/Theory_of_everything

https://en.wikipedia.org/wiki/Big_Bang

http://spark.sciencemag.org/generalrelativity/

Science News Magazine, October 17, 2015, pp. 16-31

https://www.sciencenews.org/article/lhc-restart-provides-tantalizing-hints-possible-new-particle

https://en.m.wikipedia.org/wiki/Matter

http://www.britannica.com/science/atom/The-laws-of-quantum-mechanics

http://www.nobeliefs.com/atom.htm

https://en.wikipedia.org/wiki/Physics_beyond_the_Standard_Modcl

http://plato.stanford.edu/entries/quantum-gravity/

http://dwb.unl.edu/Teacher/NSF/C04/C04Links/www.fwkc.com/encyclopedia/low/articles/q/q021000030f.html

Science in the 20th Century and Beyond 1st Edition by Jon Agar, Polity Press

http://www.britannica.com/science/

Fields of Color: The theory that escaped Einstein by Rodney A. Brooks

https://en.m.wikipedia.org/wiki/Matter

http://science.howstuffworks.com/innovation/science-questions/quantum-suicide2.htm

http://hyperphysics.phy-astr.gsu.edu/Hbase/uncer.html

https://simple.wikipedia.org/wiki/Uncertainty_principle

https://news.fnal.gov/2022/07/10-years-later-higgs-boson-discoverers-publish-refined-measurements/

https://www.space.com/higgs-boson-god-particle-explained

AUTHOR BIOGRAPHY

Near the beginning of his career with IBM Irwin Tyler revealed a diverse talent for engineering and the sciences (an earlier degree in Civil Engineering and a subsequent NY State License in Civil Engineering), for software system design and development, and for advanced research (issued a U.S. Patent in Artificial Intelligence). When IBM discovered his writing ability they provided him a platform for writing several marketing guides for the company, as well as a multimedia presentation.

It was during this period that he wrote, produced, and directed two local plays with theological themes.

After Irwin's retirement from IBM he truly found his writing calling, producing to this point twenty-three books (visit his web site **http://AhlKaynBooksandArtWork.com**), more than forty published Articles and Letters to the Editor (including Science News, Newsmax, In These Times, USAA Magazine) and has developed a growing career as a ghost writer.

His ability to make the complex understandable without compromising accuracy is a hallmark of his works. Because of his long and deep interest in the sciences and technology, he developed this business motto: "If I can research it I can write it for you." This has been his guiding principle for his previous ghost-writing clients and for his own published books, as well.

Via his Ahl Kayn Publishing company, he also edits and publishes print and digital books for other authors.

A PERSONAL NOTE

WHAT MERE WORDS CAN NOT EXPRESS

My everlasting thanks for the support and encouragement of: my wife Tzivia.

Your opinion matters:

If it moves you, please leave your thoughts about this book in a Book Review on the website from which you purchased this volume. Both Irwin and other readers would gain much from your sharing.

Thank you.

BOOKS BY Irwin Tyler (Yirmi Tyler)

END DISAPPOINTMENT BEFORE IT HAPPENS

IF I SHOULD DIE BEFORE YOU WAKE – A GUIDE TO WHAT YOU NEED TO DO NEXT

POINTS OF HEALTH The Effectiveness and Safety of Acupuncture and Acupressure

WHY ACUPUNCTURE? - When Conventional Medicine Isn't Working As You Hoped

WHY ACUPRESSURE? - When Conventional Medicine Isn't Working As You Hoped

WHY CHIROPRACTIC? - When Conventional Medicine Isn't Working As You Hoped

WHY HOMEOPATHY? - When Conventional Medicine Isn't Working As You Hoped

THE DIET CHOICE PROGRAM - Beat the Cravings and Enjoy Your Dinner

SO MANY GATES TO THE CITY... A GUIDE FOR THE MODERN PERPLEXED

A Book About Jewish Belief and Understanding, and Making Some Sense Of It

TARGUM AMERICANA – THE HEBREW BIBLE AS UNDERSTOOD BY THE SAGES

COLLECTING PAPER MONEY WITH CONFIDENCE

GRADING COINS WITH CONFIDENCE

NO TIME TO SAY GOODBYE – WHO WAS TO BLAME FOR THE DEADLIEST ACCIDENT IN AVIATION HISTORY?

MEASURING YOUR BLOOD SUGAR – FOR MEN

MEASURING YOUR BLOOD SUGAR – FOR WOMEN

AVAILABLE AT:

AMAZON.COM

LULU.COM

AHL KAYN PUBLICATIONS WEB SITE

INDEX

Age of the Enlightenment...14

alpha rays...18

antimatter...63

antineutron...63

antiproton...63

Aristotle...14

atom13, 14, 15, 16, 17, 19, 20, 21, 22, 24, 26, 27, 29, 40, 87

atomic9, 15, 41, 43, 44, 45, 53, 58, 61, 62, 63, 64, 65, 66

atomic bomb...37

atomic decay...83

atomic fission...66

atomic radiation...66

atoms...28, 29

baryons...69

Becquerel...17, 18

beta rays...18, 19

Big Bang48, 52, 53, 57, 74, 76, 79, 80

black hole...55, 71

Bohr...24, 25

boson...83

bosons...73

building blocks...69, 70

cathode ray...18

Catholic Church...14

CERN...85

charge...31

classical laws of physics...36

classical physics...33, 34, 44

Classical physics...34

closed loops...77

closed system...35

coherent superposition...43

color...21

compound...28

Conservation of Energy...35

Conservation of Mass...35, 36

Constructor Theory...78

Copenhagen interpretation..42

Curie...17

curve...50

curving...50, 56

Dalton...15

dark energy...57, 79

dark matter...57, 79

decay...82

degree of certainty..............42
Democritus12, 13, 14, 15, 67, 87
dimensions....................45, 77
Dirac.......................62, 63, 64
direction..............................40
direction of motion.............35
distortion 51, 56
$E=MC^2$..............................37
Einstein32, 33, 34, 36, 38, 39, 42, 43, 44, 48, 49, 52, 55, 56, 57, 58, 74, 75, 80
electric 31
electric charge42, 45, 47, 53, 76, 79
electricity............................47
electromagnetic33, 40, 47, 53, 65, 66, 74
electromagnetism 31, 82
electron16, 18, 19, 20, 22, 24, 25, 39, 40
electrons29, 31, 40, 45, 58, 61, 65
Electrons 38
element 28
elements.............................31
emitted light.......................24
*empty space*13, 20, 21, 23, 24, 26, 45, 80
energy9, 18, 21, 24, 25, 26, 32, 35, 36, 38, 39, 40, 41, 42, 44, 47, 48, 52, 53, 54, 55, 57, 59, 60, 61, 62, 65, 74, 79
energy level 25, 26
entanglement 71
equivalency 33
equivalent......................37, 39
expansion52, 53, 54, 57, 76, 80
expansion of the universe54, 57, 76
Fermilab.............................85
Fermions.............................71
flavors...........................68, 70
force45, 47, 49, 51, 60, 61, 62, 65, 66, 69, 71, 76
force field......................42, 45
frequency 24, 25
fundamental force 47, 76
fundamental particles70, 73, 74, 76
gas 58, 60, 61
general relativity 32
gluon..................................68
gluons 70
grand force.........................47
Grand Unified Theory....73

gravitation 75

gravitational lensing 56

graviton 71, 76

gravity31, 32, 42, 45, 47, 48, 49, 50, 53, 54, 56, 71, 74, 75, 82

hadrons 69

heat 60

Heat 31

Heisenberg 41, 45, 65

Higgs boson 82, 86

Higgs field 84

infinity 74

kinetic energy 35

laws of physics 35

leptons 69, 70, 71, 73, 80

light23, 24, 25, 31, 33, 37, 51, 55, 56, 57, 58, 65

light beam 34, 37

light wave 33

liquid 58, 60

magnetic field 42, 45

magnetic monopoles 80

magnetism 47, 53

mass30, 31, 35, 36, 37, 39, 41, 42, 44, 48, 49, 50, 52, 53, 55, 57, 59, 70, 74, 76, 80, 83, 85

masses 51, 52, 53, 54, 55

matter28, 32, 39, 41, 42, 44, 46, 47, 53, 54, 55, 57, 58, 60, 61, 63, 67, 69, 70

Maxwell 47

mesons 69, 79

Miletus 67

momentum 42, 45

motion 33, 45, 59, 60

M-Theory 76

multiple universes 43, 77

Murray Gell-Mann 68

nature.... 12, 14, 21, 24, 25, 27

negative16, 17, 22, 24, 27, 30, 45

Neils Bohr 29, 42

neutral 20, 22, 25, 26

neutrino 69, 70

neutron 23

neutrons40, 45, 65, 68, 69, 79

Newton 31

nuclear energy 37

nuclei 82

nucleus22, 23, 24, 26, 27, 29, 30, 38, 45, 61, 64, 65, 66, 68, 79, 82

observation 41, 43

observer 34, 35, 36, 42

open strings 77
open system 36
orbit........................22, 24, 25
packet 70
packets 38
particle16, 20, 22, 26, 37, 40, 41, 42, 44, 62, 65, 66, 68, 69, 71, 83, 88
particle accelerators 66
*particles*13, 15, 16, 17, 18, 19, 20, 22, 23, 24, 25, 27, 28, 38, 44, 47, 53, 58, 60, 61, 63, 64, 65, 66, 67, 68, 69, 70, 72, 73, 74, 77, 79, 80, 83
Particles 38
Pauli...................................64
photon 84
photons...................65, 66, 71
physics 10, 21, 24
pion 66, 69
Planck 21
Planck length 72
plasma................................61
plum pudding......................29
Plum Pudding...16, 17, 19, 24
point-like particle 72
position............42, 45, 55, 59
positive16, 17, 19, 20, 21, 22, 23, 24, 27, 29, 45
positron 62, 64
potential.............................42
probabilities 43
probability.....................41, 42
proton................................39
protons30, 40, 45, 58, 64, 65, 68, 79, 83
quanta 21, 100
quantum9, 10, 11, 12, 39, 41, 43, 44, 83, 100
QUANTUM 29
Quantum Field Theory48, 74, 80
quantum foam 72
Quantum mechanics 53
Quantum Mechanics 58, 74
quantum physics 10
quantum theory............10, 12
Quantum Theory 58
quantum-gravity 54, 89
quark62, 68, 69, 70, 71, 72, 79
quarks68, 69, 70, 71, 72, 73, 80
radiation.................18, 21, 25
redshift 51

relative 34
relativistic....................34, 35
Relativity44, 48, 49, 52, 54, 55, 56, 57, 75, 80
ripple 56
Rutherford17, 18, 19, 22, 23, 24
Schrödinger........................64
shape............................59, 60
singularity..........52, 53, 54, 57
solid.......................58, 60, 61
space9, 13, 20, 21, 22, 23, 24, 26, 32, 34
spacecraft...........................39
space-time49, 50, 52, 54, 55, 56, 57, 72, 75
space-time foam..................72
special relativity..................32
Special Relativity 35, 37, 38
spectrum............................31
speed......................33, 35, 40
speed of light31, 33, 35, 52, 57, 64, 80, 84
Speed of Light33, 34, 36, 37, 38, 39
Standard Model70, 73, 74, 78, 81, 82, 85
stars 46, 51, 53, 54, 61
states 43
String Theory 76, 77, 79
strong force.......66, 70, 75, 82
structure........................19, 27
subatomic 44, 53, 66
supermembrane 54
superstring..........................54
Superstring/M 76
supersymmetric matrix........54
Supersymmetry 71
theory of everything......48, 80
Theory of Everything47, 73
thermal energy 35
time32, 34, 45, 49, 50, 52, 54, 55, 57, 59, 65, 69, 74, 75, 77, 78, 79
ToE 73, 77, 78, 80
Uncertainty Principle .39, 42
unified force........................54
universe42, 43, 48, 52, 57, 63, 74, 75, 76, 77, 78, 79, 80
vapor 58, 60
velocity 39, 41, 45
vibrating........................59, 76
W and Z boson 84
W boson 85
warped.........................49, 54

warping 50, 52, 56
wave..................37, 41, 43, 83
wave-particle duality...........38
waves 44, 47, 56, 64
Waves 38
weak force.....................66, 70
weak forces 82
weak nuclear force 83
Werner Heisenberg 39
worm hole...........................77
wormhole............................54
x-rays 18, 31
Yukawa 65
Yuval Ne'man 68
Z particle 66

Made in the USA
Las Vegas, NV
25 November 2024